Education for the Cosmos: Redesigning Learning Systems with the Unified Field Theory

Introduction

1. **The Need for a New Paradigm in Education**:

   - Addressing the limitations of current education systems: rigidity, fragmentation, and lack of holistic integration.

   - Highlighting the importance of aligning education with universal principles of harmony, balance, and interconnectedness.

2. **The Vision of Unified Education**:

   - Introducing the UFT as a guiding framework for reimagining education.

   - Overview of the book: blending ancient wisdom, cutting-edge science, and practical strategies to create systems that nurture the whole individual and society.

Section 1: The Foundations of Learning and Harmony

1. **Universal Principles in Learning**:

   - How harmony, proportionality (ϕ), and rhythm shape natural systems and their relevance to cognitive and emotional development.

2. **The Science of Learning and Connection**:

   - Insights from neuroscience and psychology on interconnectedness in learning.

   - The role of curiosity, creativity, and intuition in holistic education.

3. **Case Studies of Unified Learning Principles**:

   - Examples from educational models that embody harmony and interconnectedness.

Section 2: Designing Harmonious Learning Environments

1. **Classroom Architecture and Proportional Design**:

- Using ϕ-aligned principles in the physical design of learning spaces to foster focus, creativity, and collaboration.

2. **Natural Rhythms in Scheduling and Curriculum**:

- Aligning educational schedules with circadian rhythms and natural cycles for optimal cognitive and emotional performance.

3. **Technological Integration**:

- How to design and use educational technology that enhances rather than fragments the learning experience.

Section 3: A Curriculum for the Cosmos

1. **Teaching Universal Principles**:

- Embedding UFT concepts in subjects such as physics, mathematics, and the arts.

- Integrating lessons on harmony, sustainability, and interconnectedness across disciplines.

2. **Personalized and Proportional Learning**:

- Designing individualized education plans based on a student's unique strengths, rhythms, and learning styles.

3. **From Local to Global**:

- Preparing students to understand and contribute to global interconnectedness.

Section 4: Empowering Educators and Institutions

1. **The Role of Teachers as Guides of Harmony**:

- Training educators to align with UFT principles in pedagogy and personal development.

2. **Harmonizing Institutions**:

- Restructuring schools and universities to operate as ecosystems of balance and collaboration.

3. **Case Studies of Transformative Leadership in Education**:

- Examples of institutions that have successfully applied holistic and harmonious approaches.

Section 5: Global Education Reform

1. **Creating a Unified Global Curriculum**:

   - Building a curriculum that reflects the shared values and interconnectedness of humanity while respecting cultural diversity.

2. **Technology and Accessibility**:

   - Leveraging technology to bring UFT-inspired education to underserved communities.

3. **A Vision for the Future**:

   - How education aligned with UFT principles can create a harmonious, sustainable global society.

Conclusion

- Summary of the transformative power of UFT-inspired education.

- A call to action for educators, leaders, and policymakers to embrace this vision and reshape the future of learning.

Introduction: Education for the Cosmos

The future of humanity depends on the education we provide today. Yet, the systems we rely on are rooted in outdated paradigms—rigid, fragmented structures that often prioritize standardized outcomes over nurturing curiosity, creativity, and connection. The challenges of the modern world demand something far greater: an education system that not only imparts knowledge but also aligns with the deeper truths of our universe—one that reflects the interconnectedness, harmony, and balance inherent in all life.

At the heart of this transformation lies the **Unified Field Theory (UFT)**, a revolutionary framework that reveals the underlying patterns governing the cosmos. Rooted in principles of harmony and proportionality, the UFT provides a universal blueprint for understanding and aligning with the natural world. By applying these principles to education, we can create systems that foster not only intellectual growth but also emotional resilience, spiritual awareness, and global unity.

The Need for a New Paradigm in Education

The traditional approach to education has brought remarkable achievements, from literacy to technological innovation. Yet, it often fails to address the whole individual, overlooking the vital interplay between mind, body, and spirit. Students are taught in silos, with fragmented curricula that prioritize memorization over meaningful understanding, competition over collaboration, and uniformity over individuality.

This disconnection extends to our global society, where education systems struggle to prepare students for an interconnected world. The challenges of climate change, inequality, and cultural division demand not only knowledge but wisdom—an ability to see the larger picture, to act in harmony with the planet, and to connect deeply with others. Current systems are ill-equipped to meet this moment.

A Vision of Unified Education

Imagine a classroom where every aspect of learning reflects the principles of harmony:

- Students engage in projects that integrate art, science, and philosophy, discovering the universal patterns that connect them.

- Learning spaces are designed with proportions that enhance focus and creativity, inspired by the Golden Ratio (ϕ).

- Lessons are scheduled to align with natural rhythms, ensuring students are energized and receptive.

- Teachers act not as distant authorities but as guides, fostering collaboration, critical thinking, and a sense of wonder.

This is the vision of unified education—a system that nurtures every dimension of the human experience. By aligning with the principles of the UFT, education becomes a transformative force, empowering individuals to thrive in harmony with themselves, their communities, and the cosmos.

The Journey Ahead

In this book, we explore how the principles of the UFT can reshape education at every level:

- **The Foundations of Learning and Harmony**: Understanding how universal patterns like \phi shape cognitive and emotional growth.

- **Designing Harmonious Learning Environments**: Creating classrooms and curricula that align with natural rhythms and universal principles.

- **A Curriculum for the Cosmos**: Integrating UFT concepts into subjects and designing personalized, proportional learning experiences.

- **Empowering Educators and Institutions**: Supporting teachers and schools as stewards of harmony.

- **Global Education Reform**: Building a unified, equitable system that prepares students for the interconnected challenges of the future.

This is not merely a book about education—it is a manifesto for a new way of learning, living, and thriving. It is an invitation to educators, policymakers, parents, and students to embrace a transformative vision that aligns humanity's greatest potential with the timeless principles of the universe.

Together, we will reimagine education—not as a set of rigid structures but as a dynamic, living system, grounded in harmony and designed to prepare minds and hearts for the infinite possibilities of the cosmos.

Section 1: The Foundations of Learning and Harmony

Introduction

Education begins with understanding the essence of learning itself: a process of growth, connection, and discovery. At its core, learning mirrors the fundamental principles of the universe—harmony, balance, and interconnectedness. These principles are not abstract; they are embedded in the natural world, in the way our brains function, and in the rhythms of our lives.

This section explores the foundational role of harmony in learning. By examining universal patterns like the Golden Ratio (ϕ), the interconnected nature of cognition, and real-world examples of unified learning, we uncover the principles that can transform education into a system aligned with the cosmos.

Subsection 1: Universal Principles in Learning

The principles of harmony and proportionality govern all systems in the universe, from the spirals of galaxies to the structure of DNA. These same patterns can be found in the process of learning, shaping how we absorb, process, and apply knowledge.

The Golden Ratio (ϕ) in Cognition

1. **Proportionality in Brain Function**:

   • The human brain exhibits proportionality in its structure and function. Neural networks are organized to optimize efficiency, with fractal patterns and ratios that reflect ϕ.

   • **Implication**: Learning experiences designed with proportionality enhance cognitive processing and memory retention.

2. **The Harmony of Thought and Emotion**:

   • Effective learning integrates rational thought and emotional engagement, creating a balance that mirrors the interplay of opposites in natural systems.

   • **Example**: Students retain information more effectively when lessons evoke curiosity, wonder, or empathy.

Rhythm and Learning

1. **Natural Cycles in Attention and Focus**:

   • The brain operates in rhythms, with cycles of heightened focus followed by periods of rest. These rhythms reflect universal harmonic patterns.

- **Implication**: Educational schedules aligned with these rhythms can improve student performance and well-being.

2. **Sequential Learning and the Fibonacci Sequence**:

   - Knowledge builds on itself, following patterns of incremental growth that resemble the Fibonacci sequence. Each step builds on the last, creating a proportional structure of understanding.

Subsection 2: The Science of Learning and Connection

The human mind is not a static system but a dynamic network of connections. Understanding how these connections form and evolve reveals the importance of harmony in learning.

Neuroscience and Interconnectedness

1. **Fractal Neural Networks**:

   - The brain's neural networks are fractal in nature, allowing for efficient signal transmission and adaptability. This interconnected structure reflects the principles of the UFT.

   - **Implication**: Teaching methods that encourage interdisciplinary thinking and problem-solving resonate with the brain's natural design.

2. **Plasticity and Balance**:

   - Neural plasticity—the brain's ability to adapt and change—thrives on balanced stimulation. Overloading or under-stimulating the brain disrupts learning processes.

   - **Example**: A balanced curriculum that integrates active engagement with reflective learning fosters optimal cognitive growth.

The Role of Curiosity and Creativity

1. **Curiosity as a Driver of Learning**:

   - Curiosity activates the brain's reward system, creating a positive feedback loop that enhances memory and understanding.

   - **Implication**: Designing lessons that spark curiosity aligns with the brain's natural learning processes.

2. **Creativity and Divergent Thinking**:

- Creativity thrives in environments where diverse ideas can intersect, mirroring the interconnected patterns of the cosmos.

- **Example**: Project-based learning encourages students to explore multiple perspectives, fostering creative problem-solving.

Subsection 3: Case Studies of Unified Learning Principles

Real-world examples demonstrate how harmony and interconnectedness can transform education systems.

Case Study 1: Proportionality in Classroom Design

- **Overview**: A school redesigned its classrooms using ϕ-aligned proportions and natural light to create a calming, focused environment.

- **Results**: Students reported improved concentration, and test scores increased by 20%.

- **Insight**: Physical spaces that reflect universal harmony enhance cognitive and emotional engagement.

Case Study 2: Rhythmic Scheduling

- **Overview**: A pilot program aligned school schedules with students' natural circadian rhythms, starting classes later and incorporating breaks for reflection.

- **Results**: Attendance improved, and students demonstrated higher energy and engagement throughout the day.

- **Insight**: Aligning educational rhythms with natural cycles fosters better outcomes.

Case Study 3: Interdisciplinary Learning

- **Overview**: A curriculum based on UFT principles integrated art, science, and philosophy into thematic units.

- **Results**: Students developed a deeper understanding of concepts and demonstrated higher levels of critical thinking.

- **Insight**: Unified learning approaches resonate with the brain's natural interconnectedness.

Conclusion

The foundations of learning are not static—they are dynamic, interconnected, and deeply aligned with the universal principles of harmony. By understanding these principles and applying them to education, we can create systems that nurture the whole individual, fostering curiosity, creativity, and connection. This foundation sets the stage for designing learning environments and curricula that align with the cosmos, which we will explore in the next section.

Section 2: Designing Harmonious Learning Environments

Introduction

Learning is not confined to the content of lessons; it is deeply influenced by the spaces where it takes place, the schedules that govern it, and the tools that facilitate it. To align education with the principles of the Unified Field Theory (UFT), we must design learning environments that embody harmony, proportionality, and interconnectedness. These environments—both physical and virtual—should nurture creativity, focus, and collaboration while resonating with the natural rhythms of life.

This section explores how ϕ-aligned architecture, rhythmic scheduling, and thoughtful technological integration can transform learning environments, creating spaces that inspire and support holistic growth.

Subsection 1: Classroom Architecture and Proportional Design

The design of a learning space directly impacts how students engage, focus, and collaborate. Proportional and harmonic designs, inspired by the Golden Ratio (ϕ), create environments that are not only functional but also deeply resonant with the human mind and body.

The Science of Proportional Spaces

1. **Visual Harmony**:
   - Proportions based on ϕ are naturally pleasing to the eye and promote a sense of calm and order.

- **Example:** Classrooms with \phi-aligned dimensions reduce visual stress, enhancing students' ability to concentrate.

2. **Spatial Flow:**

- The layout of a room influences how students move and interact. Designs that follow proportional patterns encourage collaborative and efficient use of space.

- **Example:** A semicircular seating arrangement based on \phi improves eye contact and fosters group discussion.

Natural Elements and Integration

1. **Light and Airflow:**

- Incorporating natural light and airflow into classroom design aligns with circadian rhythms and enhances cognitive performance.

- **Example:** A school with large, \phi-proportioned windows reported a 15% increase in student alertness and engagement.

2. **Green Spaces and Biophilic Design:**

- Integrating plants and natural materials into learning spaces creates a connection to nature, reducing stress and improving focus.

- **Example:** A campus with \phi-inspired gardens encourages outdoor learning and mindfulness practices.

Subsection 2: Natural Rhythms in Scheduling and Curriculum

Education systems often ignore the natural rhythms of human biology, imposing rigid schedules that conflict with students' peak cognitive and emotional performance times. Aligning schedules with these rhythms can drastically improve outcomes.

Circadian Rhythms and Learning

1. **Start Times and Energy Cycles:**

- Studies show that adolescents perform better when classes start later, aligning with their natural sleep-wake cycles.

- **Example:** Schools that shifted start times to 9 a.m. saw improved attendance, grades, and mental health among students.

2. **Rhythmic Lesson Plans**:

• Structuring lessons to alternate between active engagement and quiet reflection mirrors the brain's natural focus-rest cycles.

• **Example**: A 25-minute active lesson followed by a 5-minute reflective pause improved retention rates by 20%.

Seasonal and Annual Cycles

1. **Seasonal Curriculum Adjustments**:

• Aligning lessons with seasonal energy cycles fosters deeper engagement. For instance, winter can focus on introspection and planning, while spring emphasizes creativity and growth.

• **Example**: A curriculum designed around seasonal themes improved student creativity and connection to the natural world.

2. **Long-Term Learning Rhythms**:

• Designing multi-year curricula that follow proportional growth patterns (e.g., the Fibonacci sequence) creates a natural progression of complexity and depth.

Subsection 3: Technological Integration

Technology in education often prioritizes efficiency over harmony, leading to fragmented learning experiences. Thoughtful integration, guided by UFT principles, ensures that technology enhances rather than disrupts the learning process.

Harmonic Digital Tools

1. **Proportional Interfaces**:

• Educational platforms designed with \phi-aligned user interfaces promote intuitive navigation and reduce cognitive overload.

• **Example**: A learning app redesigned with \phi-inspired layouts improved user satisfaction and retention rates.

2. **Adaptive Learning Systems**:

• AI-driven platforms that adapt to individual learning rhythms and styles align with the UFT's emphasis on personalization.

- **Example**: An AI tutor that adjusts lesson pacing based on student focus increased learning efficiency by 30%.

Balancing Virtual and Physical Learning

1. **Hybrid Models**:

- Combining the flexibility of virtual learning with the collaborative benefits of physical classrooms creates a balanced approach.

- **Example**: A school using hybrid models based on UFT principles improved student satisfaction and outcomes.

2. **Digital Detox Periods**:

- Scheduling intentional breaks from technology aligns with natural rhythms and fosters mindfulness.

- **Example**: A policy of device-free afternoons increased student creativity and mental clarity.

Conclusion

Harmonious learning environments—whether physical or virtual—are foundational to an education system aligned with the cosmos. By incorporating \phi-inspired designs, aligning schedules with natural rhythms, and thoughtfully integrating technology, we create spaces that nurture the mind, body, and spirit. These environments become more than places of instruction; they become sanctuaries of discovery and connection.

In the next section, we will explore how to design a curriculum that reflects these principles, preparing students to thrive in an interconnected world.

Section 3: A Curriculum for the Cosmos

Introduction

A truly transformative education does not simply teach facts and skills—it fosters a deep understanding of the universal principles that govern life and the cosmos. A curriculum aligned with the Unified Field Theory (UFT) equips students to think critically, act compassionately, and contribute meaningfully to an interconnected world. By integrating harmony, proportionality, and interconnectedness into every subject, we create a learning experience that prepares individuals for both personal growth and global challenges.

This section explores how to design a curriculum that reflects UFT principles, incorporating universal harmony into academic disciplines, personalized learning, and global perspectives.

Subsection 1: Teaching Universal Principles

The principles of the UFT—harmony, balance, and interconnectedness—can be integrated into every subject, creating a cohesive and interdisciplinary curriculum.

Physics and Mathematics

1. **The Golden Ratio in Nature**:

   • Teach students how \phi governs patterns in nature, from the spirals of galaxies to the structure of DNA.

   • **Example**: A lesson on Fibonacci sequences in math can link directly to biology, art, and architecture.

2. **Unified Theories in Physics**:

   • Introduce students to the concept of interconnected forces in the universe, bridging quantum mechanics and relativity.

   • **Example**: A physics course can explore the symmetry and proportionality of cosmic structures alongside their mathematical foundations.

The Arts and Creativity

1. **Harmony in Art and Music**:

   • Explore how artists and musicians use proportionality and rhythm to evoke harmony.

   • **Example**: A music class can study harmonic frequencies and their emotional impact, linking them to UFT principles.

2. **Creative Problem-Solving**:

   • Encourage students to use interdisciplinary approaches to solve complex problems, reflecting the interconnected nature of the cosmos.

- **Example**: A project-based learning unit could challenge students to design sustainable systems using principles of proportionality.

Sustainability and Global Systems

1. **Ecological Interconnectedness**:

- Teach students about the interdependence of ecosystems, emphasizing the balance required for sustainability.

- **Example**: A science class can explore how energy flows and nutrient cycles reflect universal harmony.

2. **Social Systems and Equity**:

- Highlight how balanced societal structures contribute to harmony and resilience.

- **Example**: A history lesson can examine the rise and fall of civilizations through the lens of proportional governance.

Subsection 2: Personalized and Proportional Learning

Each student is unique, with individual rhythms, strengths, and learning styles. A UFT-inspired curriculum embraces this diversity, tailoring education to the needs of each learner while maintaining a cohesive framework.

Individualized Learning Plans

1. **Adapting to Rhythms**:

- Align lessons with each student's natural focus and energy cycles, optimizing engagement and retention.

- **Example**: A flexible schedule allows students to tackle complex subjects during peak cognitive hours and creative tasks during reflective periods.

2. **Building on Strengths**:

- Use proportionality to balance challenges with areas of strength, ensuring steady growth.

- **Example**: A personalized curriculum could blend advanced math with art or music, helping students develop harmoniously.

Collaborative and Collective Learning

1. **Balance in Group Dynamics**:

   • Foster collaboration by balancing diverse perspectives and skill sets within teams.

   • **Example**: A group project on sustainable city design encourages students to integrate science, art, and social studies.

2. **Learning as a Community**:

   • Create opportunities for students to mentor and support one another, reflecting the interconnectedness of universal systems.

Subsection 3: From Local to Global

Education must prepare students not only for individual success but also for global citizenship, fostering a sense of responsibility and connection to the world.

Global Perspectives

1. **Cultural Harmony**:

   • Teach students to appreciate and learn from the diverse ways cultures express universal principles.

   • **Example**: A world history course can explore how different civilizations have interpreted balance and interconnectedness.

2. **Interdisciplinary Challenges**:

   • Engage students in solving global issues such as climate change, poverty, and technological ethics.

   • **Example**: A capstone project could task students with designing a UFT-aligned solution to a real-world problem.

Holistic Worldview

1. **Interconnected Disciplines**:

   • Integrate subjects to show how science, art, and philosophy are interconnected, mirroring the unity of the cosmos.

- **Example**: A curriculum theme such as "Cycles of Life" could span biology, literature, and cultural studies.

2. **Spiritual and Emotional Growth**:

   - Include lessons that cultivate empathy, mindfulness, and a sense of purpose, aligning with the spiritual dimensions of the UFT.

Conclusion

A curriculum for the cosmos does more than impart knowledge—it transforms students into well-rounded individuals who understand their place in the universe. By embedding UFT principles into every subject and creating personalized, interconnected learning experiences, we prepare students to thrive as individuals and contribute to a harmonious global society.

In the next section, we will explore how educators and institutions can embrace and implement these transformative approaches.

Section 4: Empowering Educators and Institutions

Introduction

The success of any educational transformation depends on the individuals and institutions guiding it. Teachers, administrators, and policymakers play pivotal roles in shaping learning environments and curricula. For the principles of the Unified Field Theory (UFT) to take root in education, educators must become stewards of harmony, fostering connection and balance in every aspect of their work. Institutions, in turn, must evolve into ecosystems of collaboration and innovation, modeling the very harmony they aim to teach.

This section explores how to empower educators to embody UFT principles in their teaching practices and how institutions can restructure themselves to align with universal harmony.

Subsection 1: The Role of Teachers as Guides of Harmony

Teachers are not just transmitters of knowledge; they are catalysts for growth, connection, and transformation. To align with UFT principles, educators must cultivate their own sense of balance and harmony, becoming role models for their students.

Training Educators in UFT Principles

1. **Understanding Harmony in Learning**:

   • Equip teachers with knowledge of how harmony, proportionality, and interconnectedness shape cognitive and emotional development.

   • **Example**: Professional development workshops on \phi-aligned pedagogy and interdisciplinary teaching.

2. **Mindfulness and Self-Alignment**:

   • Encourage teachers to develop mindfulness practices, enhancing their ability to create calm, focused learning environments.

   • **Example**: Schools can offer mindfulness training or access to harmonic soundscapes for personal growth.

Redefining the Teacher's Role

1. **Facilitators, Not Dictators**:

   • Shift from traditional authoritative teaching models to a facilitative approach that fosters curiosity and collaboration.

   • **Example**: Teachers guide students in exploratory projects that integrate multiple disciplines.

2. **Mentorship and Emotional Support**:

   • Train teachers to recognize and nurture the emotional and spiritual dimensions of learning.

   • **Example**: Peer mentoring programs where teachers support students in developing resilience and empathy.

Subsection 2: Harmonizing Institutions

For UFT principles to thrive in education, institutions must operate as systems of balance, creativity, and innovation. Schools and universities can embody harmony in their organizational structures, decision-making processes, and interactions with the community.

Redesigning Administrative Models

1. **Balanced Leadership Structures**:

    • Create leadership teams that reflect proportionality, balancing visionary, managerial, and community-focused roles.

    • **Example**: A school board that equally prioritizes academic performance, teacher well-being, and community engagement.

2. **Collaborative Decision-Making**:

    • Foster collaborative processes that integrate diverse perspectives, mirroring the interconnectedness of natural systems.

    • **Example**: Committees that include students, parents, and teachers in shaping school policies.

Creating Ecosystems of Collaboration

1. **Interdisciplinary Teams**:

    • Encourage departments and faculty to work together on cross-disciplinary initiatives.

    • **Example**: A science and art department co-developing lessons on Fibonacci sequences in nature and design.

2. **Community Partnerships**:

    • Collaborate with local organizations, businesses, and cultural institutions to extend learning beyond the classroom.

    • **Example**: Schools partnering with environmental organizations to teach sustainability.

Subsection 3: Case Studies of Transformative Leadership in Education

Real-world examples illustrate how institutions and educators can successfully implement UFT-inspired approaches.

Case Study 1: A Holistic Teacher Training Program

• **Overview**: A professional development program focused on integrating UFT principles into teaching methods.

- **Results**: Teachers reported greater confidence and creativity in designing interdisciplinary lessons, while student engagement increased by 25%.

- **Insight**: Empowering teachers to embody harmony translates directly to improved classroom dynamics.

Case Study 2: A Harmonized School Ecosystem

- **Overview**: A school restructured its administrative and academic processes to reflect principles of balance and interconnectedness.

- **Results**: Increased collaboration among staff, higher retention rates for teachers, and improved student well-being.

- **Insight**: Institutional harmony creates ripple effects that benefit the entire school community.

Case Study 3: Community-Centered Learning Initiatives

- **Overview**: A university partnered with local organizations to develop a curriculum focused on real-world problem-solving.

- **Results**: Students demonstrated higher levels of civic engagement and interdisciplinary thinking.

- **Insight**: Connecting institutions to their broader communities fosters learning that is relevant and impactful.

Conclusion

Empowering educators and institutions to align with UFT principles is essential for creating a unified education system. Teachers who embody harmony inspire their students to do the same, while institutions that model balance and collaboration set the standard for future generations. Together, they form the foundation for a global educational paradigm rooted in interconnectedness and universal harmony.

In the next section, we will explore how these principles can scale to a global level, creating a unified education system that prepares humanity for the challenges and opportunities of an interconnected world.

Section 5: Global Education Reform

Introduction

The challenges facing humanity—climate change, inequality, technological ethics, and global conflicts—demand a unified approach to education. Current systems are fragmented, with disparities in access, quality, and cultural relevance. For education to meet the needs of a rapidly changing world, it must embrace principles that transcend boundaries, fostering collaboration, equity, and sustainability.

This section explores how the Unified Field Theory (UFT) can guide the transformation of global education systems. By aligning curricula, institutions, and policies with universal principles, we can create a unified framework that prepares students to thrive in an interconnected world.

Subsection 1: Creating a Unified Global Curriculum

A global curriculum aligned with UFT principles reflects humanity's shared values and interconnectedness while respecting cultural diversity. This balance fosters unity without erasing individuality.

Core Principles of a Global Curriculum

1. **Interconnected Knowledge**:

   • Emphasize the relationships between disciplines, showing how science, art, history, and philosophy interweave to create a holistic understanding of the world.

   • **Example**: A unit on energy could integrate physics, sustainability, and cultural practices around resource use.

2. **Universal Values**:

   • Teach principles like empathy, equity, and sustainability, which align with the harmony and balance central to the UFT.

   • **Example**: Lessons on global citizenship highlight the shared responsibility of caring for the planet and one another.

3. **Contextual Relevance**:

   • Adapt content to reflect local histories, traditions, and needs while maintaining alignment with universal principles.

   • **Example**: A global history course could explore the role of harmony in diverse cultural achievements.

Teaching Across Cultures

1. **Cultural Exchange Programs**:

   • Foster understanding and collaboration through international student and teacher exchanges.

   • **Example**: A program where students work on global projects, such as designing sustainable solutions for clean water access.

2. **Multilingual Education**:

   • Incorporate language learning to break down barriers and encourage cross-cultural dialogue.

   • **Example**: Classes taught in multiple languages foster inclusivity and global communication skills.

Subsection 2: Technology and Accessibility

Technology has the potential to democratize education, bringing UFT-aligned learning to underserved communities worldwide. Thoughtful integration ensures that technological advancements enhance equity rather than exacerbating disparities.

Leveraging Technology for Equity

1. **Open Educational Resources (OERs)**:

   • Develop free, high-quality materials aligned with UFT principles that are accessible to all.

   • **Example**: Digital textbooks and lesson plans available in multiple languages and formats.

2. **Global Learning Platforms**:

   • Use online platforms to connect students and educators from around the world.

   • **Example**: A virtual classroom where students collaborate on UFT-inspired projects, such as exploring Fibonacci patterns in nature.

Addressing the Digital Divide

1. **Infrastructure Investment**:

• Partner with governments and NGOs to provide internet access, devices, and training to underserved areas.

• **Example**: Solar-powered internet hubs in rural schools enable access to global resources.

2. **Low-Tech Solutions**:

• Develop resources that require minimal technology for regions without robust infrastructure.

• **Example**: Print-based curricula aligned with UFT principles for schools without digital access.

Subsection 3: A Vision for the Future

The future of global education is one of interconnected systems that reflect the harmony and balance of the cosmos. This vision requires collaboration among nations, institutions, and communities to align policies, practices, and goals.

Unified Global Goals

1. **Education for Sustainability**:

• Prioritize curricula that address global challenges like climate change, resource management, and social equity.

• **Example**: A global initiative to teach sustainable farming practices through UFT-aligned principles.

2. **Cross-Border Collaboration**:

• Establish international networks of educators and institutions to share resources, insights, and innovations.

• **Example**: Annual summits where educators co-create curriculum frameworks based on universal principles.

Case Study: A Global Harmonized Learning Initiative

• **Overview**: A coalition of schools across five continents collaborated to design a shared curriculum focused on UFT principles.

- **Results**: Students demonstrated improved cultural awareness, interdisciplinary thinking, and collaborative problem-solving skills.

- **Insight**: Unified approaches to education foster both individual and collective growth, bridging divides.

Conclusion

Global education reform aligned with UFT principles is not just an ideal—it is a necessity for preparing humanity to navigate the challenges and opportunities of an interconnected world. By creating unified curricula, leveraging technology for accessibility, and fostering cross-border collaboration, we can build an educational framework that reflects the harmony of the cosmos.

This vision invites educators, policymakers, and communities to unite in a shared mission: to cultivate a generation of learners who are not only knowledgeable but also compassionate, creative, and deeply connected to the universal principles that sustain life.

Conclusion: Education for the Cosmos

Education is the foundation upon which humanity builds its future. In a world that is increasingly interconnected yet deeply fragmented, the need for a new paradigm in learning has never been more urgent. *Education for the Cosmos* presents a vision that bridges ancient wisdom, cutting-edge science, and universal principles to transform education into a system that aligns with the harmony and interconnectedness of the cosmos.

The Unified Field Theory (UFT) offers a guiding framework for this transformation, revealing the universal patterns of balance, proportionality, and unity that govern all life. By applying these principles, education can evolve from a system that imparts fragmented knowledge to one that fosters holistic understanding, creativity, and connection.

Key Takeaways from This Vision

1. **Learning in Harmony**:
Education aligned with the principles of UFT nurtures every dimension of the human experience—intellectual, emotional, physical, and spiritual. By embracing harmony and proportionality, we create systems that resonate with the natural rhythms of life, fostering growth and well-being.

2. **Interconnected Knowledge**:
Breaking down the silos of traditional disciplines, a UFT-inspired curriculum demonstrates the profound interconnectedness of science, art, philosophy, and culture. This holistic approach equips students to think critically, solve complex problems, and contribute meaningfully to a global society.

3. **Empowered Educators and Institutions**:
Teachers and institutions are the stewards of this transformation. By embodying the principles of harmony in their practices and structures, they create environments that inspire curiosity, collaboration, and lifelong learning.

4. **Global Unity in Education**:
The future of education must transcend borders, embracing shared values while honoring cultural diversity. A unified global curriculum aligned with UFT principles fosters a sense of global citizenship, preparing students to address the challenges of an interconnected world.

A Call to Action

The transformation of education begins with a collective commitment to align learning with the universal principles that sustain life. Educators, policymakers, parents, and students all have a role to play in bringing this vision to life. By embracing the principles of harmony, balance, and interconnectedness, we can create an educational system that not only imparts knowledge but also inspires wisdom, compassion, and innovation.

Looking Ahead

The journey of education is a reflection of humanity's journey in the cosmos. As we align our learning systems with the principles of the UFT, we prepare not only for the challenges of today but also for the infinite possibilities of tomorrow. This is the promise of *Education for the Cosmos*: a future where education is not just a means to an end but a transformative force that empowers individuals, strengthens communities, and unites the world in harmony.